The Soul of Nature

or

A Spiritual View of Evolution and Creation

Weighing [the spiritual heart] represents a judgment... It is the assessment of man's being by the operation of natural laws.   R. Collin

**A Spiritual Evolution Press
Holmdel, New Jersey**

Copyright © January 2009 by Herb Cohen
All rights reserved. Permission is granted to copy or reprint portions for any noncommercial use, except they may not be posted online without permission.

ISBN 978-0-9822323-0-9

Library of Congress Control Number: 2008910852

Library of Congress subject headings: Evolution, spirituality, consciousness, creation, religion and science, etc.

First Edition Published March, 2009

1.1

Author Online at
www.aSpiritualEvolution.com

Table of Contents

	Page
Acknowledgements	i
Preface	iii
Reaching People	vi
A Condensed Version	1
The Existence of Humans as Evidence for Creation	2
Truth is not so Obvious	7
Our Distorted View of Evolution	9
Discussing Evolution without Fear	10
Dr. Francis Collins and DNA	13
No One has all the Answers Yet	14
What Controls Human Beings	16
Gerald Schroeder PhD.: Reconciling 6 Days of Creation with 15 Billion Years	19
Modern Science and Ancient Texts Agree	20
The Uncertainty of Our Origins	23

Overview of all Three Parts   25
    Our Book is based on certain
        Fundamental Findings   28

Part One: Introduction and Orientation   32
    Evolution as it is Seen Today   36
    The need for Open Discussion   38
    Human Genome Project and DNA   44
    Genetics and Fossil Studies
        Demonstrate All Life Forms
        Share a Common Source   45
    Describing the Unknown   47
    How much are we Controlled
        by DNA?   51

Part Two: Review of Part One   55
    Where Spiritual Texts and
        Big Bang Theory Merge   57
    The Question of Origins Becoming
        Open to a Renaissance Of
        Discussion by Many
        Scientists   62
    The Possibility of another Level of
        Development we call
        Conscious or Spiritual   69
    Evolution in Our Own Lifetime   78

Part Three: Review Part Two 81
    Continua and Free Will as
        Spiritual Developments 83
    Is the Evolution of Life forms
        Leading in Some Direction? 88
    Birth Continuum, Four Stages of
      Birth Evolution:
        Cell Transference
        Instinctive Reproduction
        Emotional Reproduction
        Soul Birth 92
    Evaluate Biblical Passages from
        a Modern Perspective 97
    Seven as a Sacred Number 100
    Turning the Creation Process
        Inward 101
    The Messenger Evolves 104
    Summation 105

Bibliography 107

Acknowledgements

I wish to thank my parents and my brother and sister for the love they gave me so unconditionally. Being loved gives me the courage to try to act in accord with my conscience.

If it weren't for the books and teachings of G. I. Gurdjieff, whose ideas and practical methods I am still studying, I might not have much to write about. Gurdjieff teaches that human beings have a unique role to play in the universe. Evolution involves the development of matter and life, but each human being, as an individual, has to awaken to his or her own need to evolve.

My friend Paul complained that popular media, like TV, lack genuine spiritual content. Paul continually supported and encouraged my effort to write this book, and spent many hours reading and editing my revisions. I could not have done it without Paul.

My wife, Dr. Maria Choy, pointed out two books about the evolution debate that gave my writing efforts a clear focus. My friend Dr. Michael Gaylor contributed intensely to aspects concerning humans and evolution. My son Nikko Montecillo demanded to know what my position was.

Author Patty de Llosa gave encouragement, insight and provided vital editing suggestions.

Writer Hope Ryden told me it was a good job on a difficult subject. Luba Stevens gave deep guidance and continued encouragement. Zenon Ushak arranged the first public talks of the material and often drove many miles to help me. His wife Daria was very supportive of my efforts.

Frank Sinclair encouraged me to bring Gurdjieff's ideas out into the world in the context of current issues. Maureen Donohue said the material brought many questions and left you thinking well after you put it down. Alyssa Rapp loved that it was not judgmental and let you draw your own conclusions. Bill Scheerer enlightened me on the Evangelical point of view. My sister Lee Dalvin played devil's advocate. Jim Adams made me question who my audience was. Tatiana listened, asked questions and made suggestions. My old friend Neil helped me let go of a title that was too long and encouraged me to find a new one. Ciro Tesoro listened and read many early versions.

A special note of thanks goes to Ellen Kostroff-Clive for her original painting on the cover, *"In the Beginning"*.

If this text evolves further, then perhaps consciousness will have played a part in that.

Preface

I consider myself very fortunate to have been born and raised in a time and place where I could find my way in life without much obstruction. The time was post World War II and the place was Brooklyn, New York. As a child I excelled at one or another of our street games or school sports. I wasn't the smartest or strongest, the most artistic or musical. But I had a sense of strategy when it came to games. And I had the urge to be the best at something. At the age of 15 I discovered the sport of fencing. It was perfect for me, skill was more important than strength, and fencing was an Olympic sport.

My participation in the Tokyo Olympics of 1964 was both the climax and the ending of my first major journey in life. Because of the extreme up and down experiences I went through in my inner self, I came home with the realization I didn't know who I really was. It became clear that in addition to the 'me' on the surface there were also other 'me's' hidden underneath. I felt if I didn't discover how to bring these 'me's' together I would never be able to rely on myself and never be able to feel whole.

Returning home I was searching for something, I knew I had lost my direction. With help I found my way into the Gurdjieff work. Gurdjieff's system of self development corresponded so closely with what I saw in myself as inner experiences that, to this day, more than 40 years later, I still find it an inspirational support in life.

As the significance of competitive fencing receded in my life, the influence of the Gurdjieff work, and, a few years later, Tai Chi began to fill the vacuum. Tai Chi is a Chinese form of spiritual development and exercise. I was fortunate enough to train under a great Tai Chi master, Da Liu. Now that I am semi-retired, I still teach Fencing and Tai Chi, but I have also ventured out into the turbulent waters of a debate over the origins of existence.

This debate draws me in because I see absurdities presented by two opposing points of view. Often these views, and their advocates, are extreme.

One view tells us this incredible experience we call life is nothing but an accident. The other view is that God created everything in six ordinary days, and not long ago at that. What is most perplexing to me is that I see almost no attempt at reconciliation, or any apparent search for an underlying, and far more profound truth regarding the nature of existence.

I remember Louis Armstrong's "Wonderful World" and Woody Allen's comment that "life sucks, and then you die". These notions capture the modern extremes of our common human experience. Imagine the feelings and thoughts each of these notions might inspire in you. As human beings we all struggle to make sense of our lives.

As I've struggled to make sense of my own life, and to develop myself as much as possible, I feel that the book on evolution is not yet fully written.

Does the evolution of nature provide the possibility for the conscious development of man? How does the development of consciousness underlie our earth and the universe? Are we here for a purpose or just by accident?

I don't want to go to the grave feeling that I kept my mouth shut when I should have made the effort to speak out and say what I feel needs to be said. So here it is. Let's see what happens.

Herb Cohen
Holmdel NJ
July 2008

Reaching People

The purpose of this project is to get these ideas out to as wide an audience as possible. Please contact me if you would like to coordinate a lecture or seminar based on this material.

The condensed version, on pages 1 to 24, is a convenient one hour introduction to this sensitive subject that can introduce people to our approach. It might be useful in cases where only a one hour presentation is feasible. Then, following the condensed version are three separate parts that provide more complete information. Each one of the four presentations is designed to be approximately one hour long.

The material is written with the wish not to offend anyone open to a genuine search for truth. Our hope is that all people find this material useful. We equate the unknown with the need for faith and believe that in spite of our modern technology, we are not any more intelligent or wiser than our ancestors. We suspect that neither politics, nor economics, nor technology can resolve today's worldwide problems. Only a realization of our true humanity can guide each of us individually, and eventually as a species, in the right direction.

The Soul of Nature
A Condensed Version

A search for truth often takes us to places we did not expect to go. Today, we take evolution as the starting point for a journey into who we are and what we need now.

It is quite strange that our modern industrialized civilization is dangerously polarized and divided in its views about many basic issues. Do we live in a created universe or do we exist only because of a freak accident? Our technology could destroy us or lead us to a more advanced level of civilization. We are living in a pivotal time. If the crucial elements for a positive result are not identified soon, it might be too late for our civilization. We propose here that a spiritual view of the purposes of life is exactly what is needed for mankind to evolve toward a positive outcome.

As an illustration, let's look at the current debate over evolution. Why does this issue immediately divide people into hostile, angry groups? Why is it so emotionally charged? What qualities of our human nature produce these violent reactions? What would make it possible to reduce and resolve these conflicts?

The Existence of Human Beings as Evidence for Creation

First we need to understand the significance of different levels of life. Let us think of a plant, a fish, a mammal, and a human. In each case, progressively, the degree of awareness and possibilities of experience is expanded by a full level. The relation of each of these organisms to their environment corresponds to a different experience of living, and thus each one of them experiences life from a different level. Did Buddha, Jesus, and other great spiritual teachers actually reach a level beyond ordinary human experience? This might be why many regard them as unique teachers.

On the other hand, there are those who believe we already experience all there is on the human level and regard religious and spiritual teachings as mere mythology and imaginative fantasy. Each of us must examine our unique, inner life experiences to see if we believe something of a higher level might exist. But just because spiritual possibilities may be beyond our ordinary level of experience, we must not dismiss them out of hand. We hope to show that evolution does indeed contain different levels of life, and that each level contains different possibilities.

Without doubt levels below the human level exist and perhaps levels above us also exist. We have galaxies, solar systems, suns, planets, and organic life on earth. Do we see ourselves as part of this cosmic framework or not? Given the current disharmonious relationship with our planet earth, we need to ask how becoming more harmonious as individuals and as a species might affect our relationship to the whole.

How would you characterize the differences between the most advanced animals on earth and human beings? How do you compare chimpanzees, orangutans, gorillas, monkeys, cats, dogs, horses, pigs, raccoons, elephants, dolphins, and whales with human beings?

I think you would find each of these animals much closer to each other than to humans, and by a very wide margin. I can see many human qualities that stand out to differentiate humans from all other animals. Our spoken and written languages, the development of music, our ability to construct buildings and create machines like computers and airplanes, the ability to cultivate vegetation, domesticate livestock, the wearing of clothing, the willingness to sacrifice ourselves for reasons other than survival such as for honor, belief or principle. And, of course, intentional pre-meditated warfare.

The point is humans are unique within the animal kingdom. Is this uniqueness also accidental? Does any other life form on earth exhibit qualities as different from all others as humans do? The concept of accident wears thin. First, matter from nothing, then life from non life, then humans from animals, then the possibility of spiritual humans such as Buddha and Jesus from ordinary humans, each an amazing leap, each an accident?

The greater difference of humans from other species cries out. Why, what for? Survival adaptation is the basis for the survival of the fittest. If it survives and reproduces it passes on its genes. Every species on the planet is living proof of its ability to survive. But animal level survival does not account for the finer qualities demonstrated by humans. So why would an accidental evolution select for properties unrelated to survival? This would contradict the basic premise of evolution. But what if there is a level of survival possible on a plateau or level higher than that of animals? This hypothesis of survival on a higher level might be more consistent with recognizing the familiar role of survival, while at the same time accounting for the huge difference between humans and all other animals on the planet. If humans were the only creatures capable of surviving on a level higher than the ordinary physical survival level, it might explain the vast gap in capacities between humans and all other creatures. The only other similar gap in nature is the enormous gulf between non living and living matter. So could humanity represent the beginnings of another level of life? It might also explain many spiritual statements made in the Bible and other texts such as 're-birth', 'awaken', 'You are of this world I am of another world', 'seek and ye shall find', 'know thyself' and many more.

In spite of all the false claims made by spirituality, religion, and science we do not throw away scientific findings, and we must not discard the search for spiritual meaning. It is our individual responsibility to separate the wheat from the chaff.

There has been a growing expectation in our modern technological civilization that eventually science will be able to explain everything. So those with strong religious beliefs often fear the atheistic assertions of some scientists. And some atheistic scientists fear the influence of extreme religious convictions. These fears are at the source of today's debate over evolution.

We all probably believe in a survival instinct. All living creatures will fight to defend their lives. But behaving in accordance with a survival instinct is not our only possible mode of behavior. Other behavior patterns exist such as bravery in the face of danger, nurturing and self sacrifice, compassion for the weak and sick, the thirst for knowledge, and more. Certainly the survival response came first, but the other behaviors are equally real and significant.

Evolution demonstrates increasing degrees of choice culminating in the possibility of human free will. We believe that choice and free will are the seeds for behavior that takes us beyond mere survival. But the more disharmonious and fearful we are within ourselves, the more likely we are to be limited to the more primitive, violent survival behaviors.

Truth is not so Obvious

With the rise of science and technology, many people accept the view that truth can only be proven by science and the scientific method and exclude notions that can't be proven scientifically. Many others pursue their search in the direction of the world of religious beliefs and spirituality but deny scientific findings that contradict their strongly held beliefs. But no matter what we believe, it can all be changed in an instant by unexpected circumstances and conditions. Disasters, war, the death of a loved one, and the approach of one's own death can transform one's outlook on life in a matter of moments.

Our position today is that underlying our ordinary beliefs is a more subtle, less obvious possibility for inner experiencing. This is the spiritual realm, and like a shadow world it is always partially hidden. We believe without exploring this more subtle realm within ourselves we cannot comprehend our full potential. Our understanding of "spiritual realm" is that, if it really exists, it too must be a part of our natural universe. The spiritual realm must obey natural laws, just as we recognize gravity, electricity, and atoms, also follow natural laws. If we cannot prove the existence of the spiritual realm it only means we do not yet understand enough about ourselves or our universe.

A belief in spirituality means we believe, in some way, all life is interconnected and that the source of spirituality lies in the origins of our universe. Spiritual essence is embodied in all living organisms. On the human level, unlike survival behavior, spiritual behavior does not seem to manifest automatically. The role of spiritual teaching or spiritual search is to help us develop spiritual behavior. This is the meaning of a spiritual path.

In the 1600s and 1700s a similar debate took place between science and a spiritual point of view. Whether the sun revolved around the earth or the earth revolved around the sun became an issue that seriously divided and disturbed European civilization for more than a century. Today's debates over creation or an accidental evolution lead us to the same dilemma. Shouldn't compatibility exist between a deeper scientific understanding of nature and a profound study of spiritual texts? As long as science and spirituality lead to different truths, mustn't we question the interpretations, methods and approaches used to get these divisive results?

Our Distorted View of Evolution

Darwin's theory is based on the existence of survival adaptations, which leads to the concept of "the survival of the fittest". Fossil evidence continues to be found to illustrate the adaptive changes within and between species, through time, that support Darwin's theory.

Religious people, mainly those who accept the early books of the Bible literally, oppose Darwin's theory because of the creation story in the Book of Genesis. This text states that a creator, God, made everything we see today in seven days. A sore point about evolution, for those who accept this account literally, is the proposal that both humans and apes have descended from a common ancestor. Many religious people deny this fervently, claiming our uniqueness with respect to animal life on earth and our direct connection with God.

Today a volatile gap exists between people. To reconcile opposing views is not possible unless people evolve to a higher level of understanding.

Discuss Evolution without Fear

Often we find it necessary to hide our feelings about these emotionally sensitive matters. It is so difficult to share our beliefs. Often we avoid those who disagree with us because we fear a confrontation that may make us question our own beliefs. And yet, to reach a higher level of understanding, we must make an effort to seek out common ground for discussion. We must not allow fear to be the basis of our personal beliefs.

A great debate has arisen between many religiously minded people and those atheistic scientists who support a purely physical view of evolution. Evolution cannot be properly evaluated as a whole without appreciating the full impact of emotional qualities, mental qualities, and properties of awareness. The contribution of increasing levels of consciousness cannot be considered purely physical. What we call consciousness are attributes we cannot yet fully identify, describe or measure.

When researching various atheistic accounts of evolution, we find incredible leaps in behavior described by expressions like pure chance, random, accidental, or instinctive reactions. Isn't it also likely that life experience, learning, awareness, and intelligence had a preparatory role in these monumental moments? Can we say choice or decision is entirely unrelated to life experience at any level of existence? And if consciousness did take part, why discredit it by reducing it back to reflex, DNA, instinctive or mindless action or just pure luck? To attribute a remote evolutionary event to accident is no more scientific than to say God made it so.

The appearance of nurturing in the evolutionary time line represents a new stage in the development of what life forms are capable of doing and being. For the first time creatures had to include the needs of others in their own lives. Caring for the young represents a new dimension in relationship development. A deeper connection was made with the succeeding generation. Is this any less a step than a development where independent cells for the first time began to function together as a multi-celled organism with a single destiny?

With caring for the newborn our destinies become linked with an external being. Isn't this the birth of responsibility? Could some property of consciousness, invisible to the fossil record, act as a mediator or catalyst between DNA and behavioral change? And could it be that in our present day and age, humanity now also needs a higher level of consciousness to help resolve these questions?

Consciousness is the defining attribute of life, and living matter seems to provide the medium for its full expression. But the full significance of consciousness itself has yet to be determined.

The possibility of having conscious intent is a significant result of evolution, and must not be ignored. We cannot view living organisms as having only one single level of physical existence. Could the development of increasing consciousness actually be preparing life forms for a purpose we do not yet fully understand?

Dr. Francis S. Collins and DNA

What do you believe about the origins of humanity? In his book "The Language of God" Dr. Francis S. Collins, scientist, and former head of the world wide human genome project, states that, on page 147, according to a Gallup poll of 2004, 45% of Americans believe "God created human beings, pretty much as they are now, at one time, within the last 10,000 years." This view contradicts today's scientific view of evolution and contradicts much other evidence as well.

Dr. Collins goes on to say "science's domain is to explore nature. God's domain is in the spiritual word, a realm not possible to explore with the tools and language of science... Spirit must be examined with one's heart, mind, and soul, and mind must embrace both realms. ... Science is powerless to answer such questions as: Why did the universe come into existence? What is the meaning of human existence? What happens after we die?"

No One has all the Answers Yet

Our universe began, somewhere, but we cannot say in space, since neither space nor time yet existed. The very word, "before", and even the current laws of physics lose their meaning at the beginning of creation. These are big ideas and without effort we cannot grasp them completely.

This is not very different from the response God gives in the Bible, when Moses asks for His name. God replies "I am that I am", meaning you can't know Him the same way you can know other things. God is an unknown.

So when we try to create harmony or, at the very least, communication with people who do not share our own beliefs or views, we need to admit when our own views are expressed in hostile, offensive ways that can lead to nothing but anger and violence. It is simply not good enough to assume you are right and anyone who disagrees with your views is an idiot or worse; which is basically where we are today.

It is not easy to face fear, but to fear the unknown is to fear our own evolution and our own possible development. Seek and ye shall find. In a universe where all energy originates from one source, how could there be more than one ultimate truth?

In his book "The Selfish Gene" Richard Dawkins, one of the greatest advocates of atheism says, on page 15, "At some particular point a remarkable molecule was formed by accident. It had the extraordinary property of being able to create copies of itself." How is this different from saying God created it? In this context, of evolutionary events, can we really know the true meaning of either of these terms, accident or God? So, we must be able to acknowledge the third and vital possibility, the unknown.

Dawkins says there is no need to have God in order to explain creation, and Collins says God created the universe and is behind evolution and the creation of DNA. In either case, I am certain both men would agree these are statements of belief and not fact.

What Controls Human Beings?

In his book, "The Language of God" Dr. Collins, on pages 90 to 91, refers to DNA, which is found in the earliest microscopic organisms in the same form as today, billions of years later, as something like a computer hard drive residing in the nucleus of every cell in our bodies.

So what is the role of DNA? Certainly DNA's direct role is diminished as we move up from the level of the cells to the level of the organs, and then again to the perspective of the whole creature. The influence of cell DNA, at each of these levels, becomes integrated into a larger surrounding environment of function and awareness.

The Bible story of the tower of Babel is about an attempt to reach heaven, a higher level. The project fails because of the chaos that follows from our human disharmonies. The whole idea that we could reach heaven by literally building a physical tower is contrary to a deeper understanding of the very essence of spirituality.

Just because we are strongly influenced by heredity, still, we have not lost all self control and free will, and we are not robots. I imagine there were some in Babylon who did not believe the tower would get them to heaven, and who resisted the temptation to join this folly.

There is an influence that comes from the past, but there is also an action required in the present that comes from the experience of the living creature. The possible impact upon heredity by action in the present is still in question among evolutionary biologists. The work of cellular biologist Bruce H. Lipton is bringing forth a new paradigm regarding the relationship between DNA and living creatures. Unlike the current idea that that our genetic inheritance can be affected only at conception by random gene mutations in the DNA, Lipton states that there is an interaction between the external environment and cell protein that causes changes in the living creature by turning on or off genes in the DNA. He refers to these genetic changes as adaptive mutations. See his book "The Biology of Belief". We propose that if action in the present can affect our survival, is it not possible that life changing decisions can affect heredity as well? The struggle to survive and develop continues throughout our life span and its influence must not be underestimated. A moment of life or death struggle is governed by much more than a cell's DNA. There is an entire level of consciousness that corresponds to the functional capacities of each level of living creature.

When we concentrate intensely on some emotional or mental activity, the physical body serves the intent of the higher level activity and not its own needs. The physical body is a vehicle capable of serving something higher than its own requirements. We engage in activities not at all needed by the body. We do not exist for the body alone. "Man does not live by bread alone".

In our modern industrialized society many people have lost their survival skills, and yet continue to exist. We turn angry and violent over many trivial things, or even worse, create life and death situations out of nothing worthwhile. It is not DNA that controls us, nor a mechanical, chaotic evolution that makes us what we are, but our own inner chaos and disorder, our own human disharmony and unawareness.

Reconciling 6 Days of Creation with 15 Billion Years

Gerald Schroeder provides a scientific calculation that reconciles both views. According to Schroeder both views are approximately correct. See http://www.geraldschroeder.com/age.html

The following is a synopsis of Schroeder's reconciliation calculation:

Scientists calculate it took about 15 billion years for the universe to reach its current size. This represents an expansion in space of about one million million times. Therefore, the ratio of time from around the beginning of the universe to time today is also about a million million to one. If creation took 6 days to reach the creation of Adam, that would be 6 million million days. If we divide that by 365 we get approximately 16 billion years. Both calculations are approximately equal!

The 6 days is looking forward from a creator's perspective, while from our perspective looking back into the past we have about 15 billion years. A creator's time cannot be measured by our time. A creator's time would move much more slowly than ours, our guess is perhaps a million million times more slowly.

Modern Science and Ancient Texts Agree

Everyday matter is based on the structure of atoms. Atoms are made of elementary particles and are connected to other particles by fields. Three possibilities exist for elementary particles and atoms. They are electrically positive, negative or neutral. This fundamental trinity represents the plan or the blueprint on which many laws of physics are based. The origin of these properties, or, as author Gerald Schroeder says, the origin of this wisdom, is unknown.

The origin of the Christian concept of the Holy Trinity, the Father, Son and Holy Spirit, may well be related to ancient knowledge of fundamental relationships that pervade the entire universe.

The trinity concept is also expressed in the famous Chinese Taoist symbol of Yin and Yang or the negative and positive held in balance by the neutral circle.

The fundamental trinity in modern physics comes face to face with the source of ancient religious wisdom. At the origins of its fundamental principles modern physical science does merge with spiritual teaching. In ancient times the study of science and spirit were much closer to each other than they are today. In today's materialistic society, there is separation of science from spirit. Hopefully this trend can eventually be reversed.

Our ordinary sense perception does not encompass all of reality. Science and spirituality both use special methods and techniques to explore more deeply. We must use telescopes, microscopes, positron scans, prayer, meditation and contemplation. Without discipline and special tools, we are doomed to be lost in worlds of our own fantasy, imagination, fear and ignorance. If we wish to understand, we have to open ourselves to learning to the best of our ability. This is how we prepare ourselves for our journey. Just as gravity attracts matter, preparation attracts Grace or spirit.

The Bible pictures Adam and Eve being expelled from the Garden of Eden where the trees of knowledge and life grow. Adam and Eve were not prepared. They could not absorb the knowledge yet. Even though we are created by God Himself, without our own personal experience of the struggle for survival, we are incapable of discovering our true potential. So God separated Himself from His creation. Once separated, Adam and Eve could begin to suffer God's absence. Like Cain, they had to become "restless wanderers on the earth". But they could wish to seek Him out again, "to rest in Me". Thus God prepares us for our journey and our possible return.

Quite early in the history of human civilizations we are given extraordinary examples of the fulfillment of this sacred journey. The lives of the founders of four of the world's great religions Moses, Buddha, Jesus, and Mohammed, appear together within a period of approximately 2,000 years, just a drop in the bucket of evolutionary time.

Ponder the evolution of existence. Material existence was initiated by the Big Bang, about 15 billion years ago. Cellular existence or living matter began on earth approximately 3.8 billion years ago. The appearance of Moses, Buddha, Jesus, and Mohammed marks a third astounding transformation of living matter into spiritual matter. What might this indicate for you and me?

The first two stages are undeniable. The third stage might always require something from us.

The Uncertainty of our Origins

Modern physics is finding phenomena throughout the universe where only empty space was thought to exist. There are magnetic fields within galaxies and even connecting us with other galaxies. It is looking more and more like everything in the universe is connected; that the universe might even be alive, a living universe. If the universe is a dead thing, an accidental creation would make a lot of sense. But if it is alive, we are certainly tiny individual parts of this great whole.

Geneticists have mapped the human Genome. The findings of common genetics leave little room to doubt the relatedness of all species, all life on earth. In "The Language of God" Francis Collins states, on page 126, that genetic diversity among humans is from 10 to 50 times less than the genetic diversity between almost any other species on earth; "Thus, by DNA analysis, we humans are truly part of one family." Population geneticists conclude that all members of our species descended from a common set of founders, approximately 10,000 in number, who lived about 100,000 to 150,000 years ago."

Controversy will continue, but a new understanding of a fully integrated, living universe is emerging. It is finally undeniable that human beings also arose as an integrated part of the whole spectrum of life. But the final word has not been heard yet. We cannot create a single living cell from raw materials. Much remains to be discovered.

Just as we know the cells in our bodies are connected to form a human being, we humans are cells in the tapestry of organic life on earth. The living earth is a cell in a cosmic framework we are only beginning to comprehend.

The Soul of Nature
Overview of All Three Parts

Part 1: Presents evidence from Genetics, DNA studies, Fossil studies and Physics, that point to the existence of a spiritual realm. We discuss the unknown, the continuum of matter and energy that makes up our universe. We discuss the concept of different levels in the universe. We end with the question of just how much are we controlled by heredity and genetics.

Part 2: The Big Bang theory in Physics currently claims everything in existence comes from the same source. DNA studies support the view that all life on earth contains the same basic DNA molecule. We discuss how the development of harmony within ourselves during our own lifetime can lead us to higher levels of consciousness.

Part 3: Science is now discussing the view that our universe might be a living organism. We illustrate and give examples of how to recognize and actually experience the spiritual realm.

A search for truth often takes us to places we did not expect to go. Today, we take the debate about evolution as a starting point for a journey into who we are and what we need now.

It is quite strange that our modern industrialized civilization is dangerously polarized and divided in its views about many basic issues. Do we live in a created universe or do we exist only because of a freak accident? What motivates science to investigate things? What motivates human beings to improve things? We are living in a pivotal time. Our technology could destroy us or lead us to a more advanced level of civilization. If the elements crucial to a positive result are not identified soon, it might be too late for our civilization. We are proposing that a spiritual view of the purposes of life is exactly what we need to put things in a perspective that could lead to a positive, productive outcome.

As an illustration, let's talk about the current debate over evolution. This issue immediately divides people into hostile, angry groups. Why is this issue so emotionally charged? What qualities of our human nature produce these violent reactions? What would make it possible to reduce and resolve these conflicts?

We all probably believe in a survival instinct. All living creatures will fight to defend their lives.

But behaving in accordance with a survival instinct is not our only possible mode of behavior. Other behavior patterns exist such as bravery in the face of danger, nurturing and self sacrifice, compassion for the weak and sick, the need to cooperate with others, and more. Certainly the survival response came first, but the other, more advanced behaviors are equally real and significant.

Evolution has provided us with capabilities different from other creatures, since we seem able to act on impulses of a far more profound nature. Perhaps evolution has seeded human beings with the possibility of a free will and a different potential for consciousness.

The more disharmonious and fearful we are within ourselves, the more likely we are to be limited to the more primitive, violent survival behaviors. Is what we believe the most important factor in determining how we act? How do we arrive at our beliefs? What forces truly guide our actions? Our book is aimed at using both current knowledge and traditional beliefs to help us find a deeper and more satisfactory approach to these questions.

Our Book is based on Certain Fundamental Findings

The latest research reveals two powerful conclusions:

- The physics of the Big Bang Theory suggests that everything in existence began from one and the same source.

- Genetics and fossil studies demonstrate that DNA directly links all life forms with previously existing life forms.

The likelihood that the universe was not accidentally produced by random forces is now being openly discussed by many scientists.

Two scientists supporting a spiritual view of creation will be highlighted in our book. Francis Collins M.D., Ph.D. is a Physician and Geneticist noted for his landmark discoveries of disease genes, as well as his leadership of the Human Genome Project (HGP).

Gerald Schroeder is the noted Ph.D. of Applied Physics in Nuclear Science of M.I.T.

We see that evolution reveals a spiritualizing process at work.

- The sequence of life form development, from one celled organisms, to plants and sea creatures, followed by land animals leading to human beings, illustrates a continual path of increased freedom of movement, increased emotional capacity and, especially with human beings, a tremendous growth of mental development with the possibility of another level of development we call conscious or spiritual.

- There exist continua in nature and especially on the human level. For example, on the physical level, there is heightened sensory development as seen in musicians, artists and athletes. On the emotional level, there is nurturing and the awareness of the needs of others, as with

Doctors, nurses, clergy and social workers. On the mental level, there is the development of inquiry into the world around us, as we see with scientists and mathematicians. In spiritual development the universe around us becomes related to our own private inner world. By searching in both the spiritual and practical realms, our knowledge and free will are evoked and further development becomes possible.

Discrepancies between scientific theory and Biblical interpretation should trigger a need to question when to interpret literally, when to look for analogy, metaphor and symbolic meaning in Biblical passages, as well as when to question the conclusions attached to scientific results. Respecting scientific evidence and honoring the spiritual perspective of the Bible and other sacred texts, is a reliable way to avoid narrow minded and ritualistic interpretations detrimental to the genuine spiritual growth of all.

Atheism needs to include in its world-view more than mental logic based entirely on ordinary sensory perception. Atheism is also associated with a fear based denial of experiences the ordinary mind is unable to classify. Isn't the existence of living organisms evidence for a world of order within chaos? Indeed, if the evolution of nature is not purely accidental and might show signs of a direction, then our job is to contemplate and understand what this might mean for us before concluding that it has no meaning at all. We do not believe present day science is sufficiently developed to warrant a scientific conclusion regarding the origins of existence.

The Soul of Nature
Part One

Introduction and Orientation

At one time or another we all question what life and our universe is about. The search to discover the sense and aim of our existence can last, for some of us, only a few moments, while for others, for almost a complete lifetime.

With the rise of science and technology, many people believe that truth can only be proven by science and the scientific method and exclude notions that can't be proven scientifically. Many others pursue their search in the direction of the world of religious beliefs and spirituality but deny scientific findings that contradict their strongly held beliefs. A third group believes it knows all they need to know about both religion and science, and prefers to spend their time on more practical matters. Certainly what people believe, and what seems important to us can be changed in an instant by unexpected circumstances and conditions. Disasters, war, the death of a loved one, can transform one's outlook on life in a matter of moments.

Our position today is that underlying our ordinary perceptions of life is a more subtle, less obvious possibility for inner experiencing. This is the spiritual realm, and like a shadow world it is always partially hidden. We believe without exploring this more subtle realm we cannot comprehend our full possibilities. Our understanding of "spiritual realm" is that, if it really exists, it too must be a natural part of our universe. The spiritual realm must obey natural laws, just as we recognize gravity, electricity, and atoms, also follow natural laws. If we cannot prove the existence of the spiritual realm it means we do not yet understand enough about ourselves or our universe. Traditional creation myths have always reflected our incomplete understanding. It was understood that creation myths were abstract descriptions and not meant to be taken literally.

Today we know all matter in the universe is in movement. All living things go through the processes of conception, gestation, childhood, maturity, old age, death and disintegration. Every living creature must learn, adapt and struggle to survive. Belief in spirituality means that life is not simply a manifestation of our physical being and instincts, that we are connected to a whole greater than our individual selves and that we must aim to understand what this means. The source of our spirituality lies in the origins of our universe. Spiritual essence is embodied in living organisms, but on the human level, unlike survival, spirituality does not seem to manifest automatically. The role of spiritual teaching is to help us develop sensitivity for the spiritual. This is the meaning of a spiritual path.

Today, there is a great rift between people arising from the apparent discrepancy between The Big Bang of 15 billion years ago or a creation in 7 days that took place less than 15,000 years ago. In Part Two we will present a calculation that proposes to reconcile this division. In the 1600's and 1700's similar debates took place between science and a spiritual point of view about whether the sun revolved around the earth or the earth revolved around the sun. This issue seriously divided and disturbed European civilization for more than a century. Today's debates over creation or an accidental evolution lead us to the same dilemma. Shouldn't compatibility exist between a deeper scientific understanding of nature and a profound study of spiritual texts? As long as science and spirituality lead to different truths, mustn't we question the interpretations, methods and approaches used to get these divisive results?

Our text is designed to diffuse and reduce the lines of division between opposing groups and allow them to come together. Those individuals who try to sustain a sincere search for truth will be the most influential in bringing about this needed reconciliation. We seek out these people and address our book to them.

Evolution as It Is Seen Today

Darwin's Theory of Evolution has stirred up an on-going debate that has lasted for 150 years and is still going strong. The crux of the debate is about whether a God or Creator exists. For many, Darwin's Theory of Evolution is assumed to explain how life and humanity came into existence. It is further assumed that if the processes of nature can explain it all, then there is no need to imagine a God to do the creating. Today we will be taking a very fresh look at these assumptions.

Darwin's theory is based on the existence of survival adaptations, which leads to the concept of "the survival of the fittest". Fossil evidence continues to be found to illustrate the adaptive changes within and between species, through time, that support Darwin's theory.

Religious people, mainly those who accept the early books of the Bible literally, oppose Darwin's theory because of the creation story in the Book of Genesis. This text states that a creator, God, made everything we see today in seven days. A sore point about evolution, for those who accept this account literally, is the proposal that both humans and apes have descended from a common primate ancestor. Many religious people deny this fervently, claiming our uniqueness with respect to animal life on earth and our direct connection with God.

The rising hostility between Islamic terrorism and the West is also directly related to the images of atheism and hedonism being projected worldwide by western media. There is hostility, violence, and enmity between those who accept a literal interpretation of the Bible and those who believe scientific findings contradict the Bible. Christian, Muslim, and other religious fundamentalists, as well as atheists, all have hardened their respective views. The debate is raging, not just in the United States, but all over the world. Is God or the spirit realm real?

We hope a deeper, more spiritual view of evolution can provide common ground for genuine and open communication between opposing groups. There is need for a reconciling view that sheds new light on our debates. We need to create an atmosphere in which opposing views can come together, and bring about a higher level of understanding that will harmonize and satisfy all groups with a new recognition of truth and facts. This would be much better than the wasteful, closed minded antagonism that currently exists. Without reaching a higher level of understanding, reconciliation will not be possible.

The Need for Open Discussion

Often we find it necessary to hide our feelings about these matters. It is so difficult to share our beliefs. We avoid those who disagree with us because we fear a confrontation that may make us question our own beliefs. And yet, to reach a higher level of understanding, we must make an effort to seek out common ground for discussion. We must not allow fear to be the basis of our personal beliefs.

Every living creature that ever existed has had to grow, adapt, learn, and struggle in order to survive, only in the end to give up everything, and die. This is the inevitable law of life. Is it so because the universe is in constant movement? Why does life repeat? Where does life energy go? Life itself and the experience of living and struggling to survive is precisely what we all share.

Ironically, our daily lives often pass without perception of these changes. Once we become adapted to our daily circumstances, life may seem static and only catastrophes reveal the dynamic forces always at work. So what is the birth and death of stars and living organisms all about? Couldn't we have stayed in the Garden of Eden and lived eternally? Is there any purpose to living and dying? What could evolution be all about?

The average person's perception of evolution is one that's driven by a blind and completely unconscious mechanical instinct to survive. In this view, there is no conscious purpose or purpose to consciousness. But people who believe in something beyond the mechanical nature of things react negatively to a view of evolution that excludes human spirituality. A great division and debate has arisen between many spiritually minded people and those scientists with an atheistic view who support a purely physical view of evolution. Evolution cannot be properly evaluated as a whole without appreciating the full impact of emotional qualities, mental qualities, and properties of awareness.

The more we contemplate evolution the more we realize that, assuming there is a creative force, evolution, or nature, is the very means through which the creative force attains its aims. Supporting evidence therefore, should reveal a pattern of development that describes a future evolutionary direction. We believe the development of consciousness in life forms through time reveals just such a pattern. The appearance of increasing levels of consciousness need to be noted when describing the results of evolution.

Even though today's world is filled with violence, chaos, and hostility, the difficulty of attaining greater harmony among people does not mean it's impossible. And wouldn't the development of genuine spirituality be the very key to such a needed striving? Or should working toward a more harmonious and peaceful world simply be dismissed as not possible, which might make sense if we assume the world was created by accident. Is the atheist's view better, as inferred by physicist Victor J. Stenger throughout his book "God: the Failed Hypothesis", that current human behavior is exactly what it should be and, if it ever changes, it will be by accident?

When researching various atheistic accounts of evolution, one finds expressions like pure chance, random, accidental, or instinctive reaction, at just those evolutionary moments manifesting incredible leaps in behaviors. Yet we are looking back on the processes of evolution, we know the results and can easily minimize the significance of changes. So those who begin with the conviction of a random, accidental universe always seem to have a very simple answer at their disposal, accident. They don't want to provide room for the unknown. We must not be afraid to show humility. Attributing a remote evolutionary event to accident is no more scientific than to say God made it so.

The first creatures that actually made breakthrough changes, such as a live birth, a concern for the survival of newborns, or learning to utilize fire, were probably not aware of the evolutionary significance of their actions. But isn't it also likely that life experience, learning, awareness, and intelligence had a preparatory role in these monumental moments? And if consciousness did take part, why reduce it all to reflex, DNA, instinctive, mindless action and just pure luck? Is choice or decision entirely unrelated to life experience at any level of existence?

Geneticists say one of several results of the development of live birthing was to give the young a better chance for survival in colder climates. No doubt it had significant survival value. The appearance of nurturing in the evolutionary time line represents a new stage in the development of what life forms are capable of doing and being. So it is not just the mechanism of birthing but the emotional capacity to care for others that matters.

Even if we cannot say exactly what it was, some kind of consciousness, some kind of emotional intent that does not seem to have existed before began to appear.

With caring for the newborn into adulthood, our destinies become linked with an external being. Isn't this the birth of responsibility? And couldn't responsibility be a higher level manifestation of the same kind of symbiosis that took place between single celled creatures and the first multi-celled organisms? Is it further possible that during the two billion years when only single celled organisms are found in the fossil record, something invisible within these one-celled creatures was developing? And could that something be some form of consciousness? And even further could invisible consciousness be the missing factor in explaining why species to new species changes found in fossil studies seems to appear in steps rather than gradually, as was expected by some? Could some property of consciousness, invisible in the fossil record, act as a mediator or catalyst between DNA and behavioral change? Could this also have influenced what some paleontologists believe to be the explosion of life forms that took place during the Cambrian period? Could these monumental changes have taken place in the complete absence of consciousness? It would be like turning mammals back into stones or reptiles. So could it also be that in our present day and age, humanity now also needs a higher level of consciousness in order to be able to deal with many of its current developmental problems?

How many of us have experienced things in life we cannot explain? Who has had strange and unique moments of inner experience? Even if these moments are fleeting and never return, they give us glimpses of something that helps us to wonder, question and seek. Consciousness is the defining attribute of life, and living matter seems to provide the medium for its full expression. But the full significance of consciousness itself has yet to be determined.

The possibility of having conscious intent is a significant result of evolution, and must not be ignored. We cannot view living organisms as having only one single level of physical existence. Could the development toward increased consciousness actually be preparing life forms for a purpose we do not yet fully understand?

Human Genome Project and DNA

What do you believe about the origins of humanity? In his book "The Language of God", Dr. Francis S. Collins, scientist, and former head of the world wide human genome project, on page 147 states that according to a Gallup poll of 2004, 45% of Americans believe "God created human beings, pretty much as they are now, at one time, within the last 10,000 years." This view contradicts today's scientific view of evolution and contradicts much other evidence as well.

Bible literalists feel scientists believing in evolution are atheists. Many scientists regard Bible literalists as people who deliberately ignore the facts. Our book will reveal a reconciling view that actually integrates both evolutionary and spiritual forces within the workings of nature.

Genetics and Fossil Studies Demonstrate That All Life Forms Share a Common Source

During the presentation ceremonies at the White House, in June, 2000, honoring the Human Genome Project for its achievement of successfully mapping the entire human genome, it was said, "Today we're learning the language in which God created life". Dr. Collins was deeply touched, and as he says on page 3 of his book "The Language of God", he felt the occasion was a call for both "… a stunning scientific achievement and an occasion for worship". The whole point of his book is to show that there is not only compatibility, but also harmony, between the scientific and spiritual world views. He goes on to say "science's domain is to explore nature. God's domain is in the spiritual word, a realm not possible to explore with the tools and language of science… Spirit must be examined with one's heart, mind, and soul, and mind must embrace both realms. … Science is powerless to answer such questions as: Why did the universe come into existence?   What is the meaning of human existence? What happens after we die?" We add, when a mind embraces both science and spirit, the two realms become part of a greater whole.

So we might begin by asking what Dr. Collins, and many others including ourselves, see as scientific evidence supporting a spiritual view of evolution. The first fact is that science does not know why the Big Bang occurred.

Describing the Unknown

In our everyday world, we can usually explain what we see. We observe the cause and the effect. But in the case of the Big Bang, we are not dealing with our usual every day world. Based on our latest scientific measurements, a universe expanding in all directions is revealed. And that expansion is accelerating. Our universe began, somewhere, but we cannot say in space, since neither space nor time yet existed. So, when we say science does not know what caused the universe to appear, we mean that according to our own theories we don't know what created the Big Bang. The word, "before", and even the laws of physics move beyond our understanding at the beginning of creation. These are big ideas and without effort we cannot grasp them completely.

We should remember the response God gives in the Bible when Moses asks for His name. God replies "I am that I am", meaning you can't know Him the same way you know other things.

The overwhelming evidence provided by studying DNA in many different species of animals, including humans, shows all animal life, indeed all life on earth, is related and has a common source. In the earliest microscopic organisms found, DNA exists in the same form as today, billions of years later. Science cannot explain the origin of this exceedingly complex molecule. It seems to have arisen out of nowhere, and yet governs the hereditary influence of all creatures then and now.

In our everyday world answers to the unknown are imperative for survival. But we must accept that investigating the origins of life and the origins of the universe are not everyday matters, and cannot have ordinary answers. In this realm, it is essential to remain modest in one's claims. Terms such as faith, belief, God, the unknown, probability and accident, require a large degree of latitude.

If we wish to create harmony or, at the very least, communication with people who do not share our own views, we need to admit when our own views are expressed in hostile, offensive ways that can lead to nothing but anger and violence. It is simply not good enough to assume you are right and anyone who disagrees with you is an idiot or worse; and that is basically where we are today.

It is not easy to face fear, but to fear the unknown is to fear our own evolution and our own possible development. Seek and ye shall find. All honest human beings, whether they are atheists or religious believers, should be able to share a sincere search for truth. In a universe where all energy originates from one source, how could there be more than one ultimate truth?

On page 15 of his book "The Selfish Gene", Richard Dawkins, one of the greatest advocates of atheism says, "At some particular point a remarkable molecule was formed by accident. It had the extraordinary property of being able to create copies of itself." How is this different from saying God created it? In this context of evolutionary events, can we really know the true meaning of either of these terms, accident or God? So, we must be able to acknowledge a third and vital possibility, the unknown.

Dawkins says there is no need to have God in order to explain creation, and Collins says God created the universe and is behind evolution and the creation of DNA. In either case, I am certain both men would agree these are statements of belief. What we believe should be a manifestation of our free will. It should not be a view that reflects only what others think. We need to take the trouble to make our choices truly our own. And room should be left for including the unknown, recognizing that, with new knowledge, what is unknown also changes.

Underlying the incredible diversity we see today, organic life on earth represents one single evolving biological system. This life is evolving within the earth's ecosystem, within our solar system, within our galaxy, and within the universe. This is taking place, whether we are aware of it or not. Yet what these relationships mean is still unknown.

How Much Are We Controlled by DNA

On page 91 of his book, "The Language of God", Dr. Collins refers to DNA as something like a computer hard drive, residing in the nucleus of every cell in our bodies. So what is the role of DNA? Certainly DNA's direct role is diminished as we move up from the level of the cells to the level of the organs, and then again to the perspective of the whole creature. At each of these levels the influence of cell DNA becomes integrated into a larger surrounding environment of function and awareness. We can see this in our own bodies. Can any single cell, with its DNA, be responsible for our desire to have ice cream one particular evening?

Behavior at the human level is quite complex. At the level of the organs, each organ has to function according to influences above its own level as well as influences below its own level. Each cell itself may not be aware of the needs of the whole organ. The forces controlling the organ are, in many cases, greater than those controlling the cell. The whole creature requires yet another level of influences. And there are still higher levels as well involving the conditions of our planet and beyond.

The Bible story of the tower of Babel is about humanity's attempt to reach the higher level of heaven. The project fails because of the chaos that follows from our human disharmonies. The whole idea that we could reach heaven by literally building a physical tower is contrary to a deeper understanding of the essence of spirituality. But many were foolish enough to try it anyway.

So what does exercise the greatest influence over the life of each living creature? The important point here is that just because we are strongly influenced, and perhaps even governed by heredity, still, we have not lost all self control and free will, and we are not robots. I imagine there were some in Babylon who did not believe the tower would get them to heaven and who resisted the temptation to join this folly.

The life of all creatures, especially humans, is determined by the interactions of the individual being with its environment. There is an influence that comes from the past, but there is also an action required in the present that comes from the experience of the living creature. The possible impact upon heredity by action in the present is still in question among evolutionary biologists. The work of cellular biologist Bruce H. Lipton is bringing forth a new paradigm regarding the relationship between DNA and living creatures. Unlike the current idea that DNA, and thus our genes, can only be affected at conception by random gene mutations in the DNA, Lipton states that there is an interaction between the external environment and cell protein that causes changes in the living creature by turning on or off genes in the DNA. He refers to these genetic changes as adaptive mutations. See his book "The Biology of Belief". We propose that if action in the present can affect our survival, is it not possible that life changing decisions can affect heredity as well? We must not underestimate the struggle to survive. A moment of life or death struggle is governed by much more than the cell's DNA. There is an entire level of consciousness that corresponds to each level of functionality possible for each living creature.

When we concentrate intensely on some emotional or mental activity, the physical body serves the intent of the activity and not its own needs. The physical body is capable of serving something higher than just its own requirements. And this is probably truer for humans than any other creature. Each whole activity can encompass several different levels simultaneously.

The survival instinct brings every drop of physical, emotional, mental, and spiritual capabilities into the moment. The unification of all faculties into one is a spiritual process. After years of neglect, the possibility to pull one's self together may become lost. Those modern civilized humans, who live everyday concerned only about themselves, and who become dependent on our many creature comforts, are often caught unprepared by power outages and other kinds of disasters or dangers. Many people have lost their survival skills, and yet continue to exist. We turn angry and violent over many trivial things, or even worse, create life and death situations out of nothing worthwhile. It is not DNA that controls us, nor a mechanical, chaotic evolution that makes us what we are, but our own inner chaos and disorder, our own human disharmony and unawareness. Growth of consciousness depends on the active impressions of our perceived environment. Striving toward inner harmony accelerates growth toward higher development.

Review of Part One

In Part One we presented evidence from the Big Bang Theory, fossil studies and Genetics to illustrate the feasibility of our universe being a creation rather than an accident. If our universe is an intentional creation, then the existence of a spiritual realm would be quite naturally implied.

In Part Two, we will present examples of a spiritual realm.

In Part Three, we will try to get a taste of what experiencing the spiritual realm is like.

Main points:

The Big Bang Theory brings the concept of a unity of unimaginable condensed energy existing at the beginning. The Big Bang Theory makes no claim to understand why the universe came into existence. However, the amazing thing is that all matter and energy in existence today arose from one source, whatever that was. There is a unity of all material existence, and this is something to be pondered.

Genetic evidence states that all living matter on Earth is related by the DNA molecule. The DNA molecule is another unity that exists in every cell that has ever lived on the earth. From this one molecule, all subsequent life forms arose. This is quite amazing. The Big Bang took place before time and space existed, so we can envision this event as the origin of our unique space and time. But picture the primeval Earth. Time and space already existed. Why did only one cell with one kind of DNA become the only one to survive? The overwhelming scientific evidence is that all life on earth arose from the same source.

Part One suggests that, as life forms become more and more complex, the influence of heredity and DNA diminishes. We humans cannot blame our behavior on DNA and heredity. We need to take responsibility for our own disharmonies and lack of awareness. If we can become more aware and reach higher degrees and levels of consciousness, then possibly our behaviors will also grow in the direction of reconciliation, and harmony will become more possible within us and with others.

The Soul of Nature
Part Two

Where Spiritual Texts and Big Bang Theory Merge

Gerald L. Schroeder begins his book "The Hidden Face of God" by saying, "A single consciousness, an all embracing wisdom, pervades the universe". He also wrote the books, "Genesis and the Big Bang" and "The Science of God". Schroeder has a PhD from M.I.T. and is also a devout student of Kabala, which is a Jewish tradition of study into the esoteric or deep meanings behind Biblical text.

The accepted translation of Genesis 1.1, "In the beginning God created the heavens and the earth..." according to Schroeder's study of Kabala, should read, "With wisdom God created the heavens and the earth..." The essence of his books is that all the latest evidence and theories in physics today suggest overwhelmingly that our universe was not the result of accidental forces.

Everyday matter is based on the structure of atoms. Atoms are made of elementary particles and are connected to other particles by fields. Three possibilities exist for particles and atoms.

They are electrically positive, negative or neutral. This fundamental trinity represents the plan or the blueprint on which many laws of physics are based. The way these forces interact determines how matter combines and develops. The origin of these properties, or this wisdom, is unknown.

The fundamental precept of modern physical science, a search for truth, does merge in principle with genuine spiritual teaching. Yet, in very ancient times, the study of science and spirit were much closer to each other than they are today.

The origin of the Christian concept of the Holy Trinity, the Father, Son and Holy Spirit, may well be related to ancient knowledge of fundamental relationships that pervade the entire universe. The trinity concept is also expressed in the famous Chinese Taoist symbol of Yin and Yang or the negative and positive held in balance by the neutral circle. The fundamental trinity in modern physics comes face to face with the source of ancient religious wisdom.

Computer bits are on or off. A computer is a wonderful device, but it is missing an essential ingredient. The computer is truly mechanical and operates with only two forces. Nature was created with three. All of nature's evolving processes require the harmonious blending of three forces to create something new.

Our ordinary sense perception does not encompass all of reality. Special methods and techniques are used by science and spirituality to explore more deeply. We must use telescopes, microscopes, positron scans, prayer, meditation and contemplation. Without special tools, we are doomed to be lost in worlds of our own fantasy, imagination, fear and ignorance. Exploring the outer world is just as difficult as exploring our own inner worlds. If we abandon either realm we inevitably lose sight of the wholeness of reality. We don't have to be scientists to appreciate physics, and we don't have to be priests to approach God. If we wish to understand, we have to open ourselves to learning to the best of our ability. This is how we prepare ourselves for our journey. Just as gravity attracts matter, preparation attracts Grace or spirit.

So, with wisdom, a creative force set things in motion. The aim is to create. As above, so below. One set of laws gives rise to the creation of physical matter, and eventually, living matter evolves. Finally, the spiritual matter that was there from the beginning also begins to evolve within living matter. Gradually, little by little, matter is evolving so that we can comprehend, can begin to become more conscious of the whole of existence. We can serve something greater than our own desires. Thus far human beings represent the greatest achievement of evolution. Has evolution brought us this far only to wage war and hate others of our own kind? The miracle of evolution for humans seems to include a unique possibility for individual development. Human beings, only a few at first, begin to realize this, and begin to seek out what this means.

    The Bible pictures Adam and Eve being expelled from the Garden of Eden, where the trees of knowledge and life grow. Adam and Eve were not prepared. They could not absorb the knowledge yet. The Bible says we are created by God Himself. Yet without our own personal experience of the struggle for survival, we are incapable of discovering our true potential. So God separated Himself from His creation. Once separated, Adam and Eve could begin to suffer God's absence. Like Cain, they had to become "restless wanderers on the earth". But they could wish to seek Him out again, "to rest in Me". Thus God prepares us for our journey and our possible return to the source again.

    Quite early in the history of human civilizations we are given extraordinary examples of this sacred journey. The founders of four of the world's great religions, Moses, Buddha, Jesus, and Mohammed, all appear within a period of approximately 2,000 years, just a drop in the bucket of evolutionary time.

Ponder the evolution of existence. Material existence was initiated by the Big Bang, about 15 billion years ago. Cellular existence or living matter began on earth approximately 3.8 billion years ago. The appearance of Moses, Buddha, Jesus, and Mohammed marks a third astounding transformation of living matter into spiritual matter. What might this indicate for you and me?

The first two stages are undeniable. The third stage might always require something from us.

The Question of Origins Is Now Open to a Renaissance of Discussion by many Scientists

Both Francis Collins and Gerald Schroeder agree the last 20 years of scientific research has witnessed tremendous breakthroughs as well as dramatic dead ends. Physicists expected to find answers that would lead to a Unified Field Theory for which Albert Einstein had been searching. Instead they found more questions, more unknowns.

Modern physics is finding phenomena throughout the universe where only empty space was thought to exist. There are magnetic fields within galaxies and even connecting galaxies. It is looking more and more like everything in the universe is connected, that the universe might even be alive, a living universe.

Paleontologists filled in many gaps in the fossil record. Compelling evidence of transitional species has been found. Throughout his book Collins supports the view that transitional fossils do indeed exist. He says, on page 147 of "The Language of God", "... virtually all of the findings are consistent with the concept of a tree of life of related organisms. Good evidence exists for transitional forms from reptiles to birds, and from reptiles to mammals". Schroeder, on pages 36 to 40 in "The Science of God", relates the story of Charles Doolittle Walcott, Director of the Smithsonian Institute. Walcott's collection of over 60,000 fossils were hidden away in the Smithsonian in 1909 and only rediscovered in the mid-1980s. They changed the concept of evolution. Prior to this rediscovery, evolutionary theory proposed that over 100 million years would be necessary to allow complex crustaceans to evolve from the one-celled protozoan that existed in the pre-Cambrian period. These Burgess fossils are overwhelming evidence for an explosion of new complex life forms taking place within a mere 20 million years. Schroeder says "The overwhelming weight of evidence tells us something exotic certainly happened to produce the variety of life as we know it on our planet. As to what that was, the jury is still out." He goes on to say "... it is our understanding of the development of life that must be revolutionized. ... He continues: "I must clarify an important point. The Burgess fossils do not question the development of classes of life. It is no secret that each individual phylum first appeared as simple aquatic forms and became more complex with the passage of time". What he is saying is that modern science still lacks a complete understanding of the mechanisms underlying the processes of evolution.

In spite of this undeniable validation of the natural process of development in the evolution of life great controversy still continues to rage with respect to the role of evolution. Some still maintain that no real transitional fossil evidence exists. We suggest reading: <u>"Wonderful Life: The Burgess Shale and the Nature of History" by Stephen J. Gould</u>.

Geneticists have mapped the human Genome. The findings of common genetic patterns within the genes of widely different species leaves little room to doubt the relatedness of all species, all life on earth. Collins, on page 126 of "The Language of God" states, "This remarkably low genetic diversity distinguishes us from most other species on the planet, where the amount of genetic diversity is ten or sometimes even fifty times greater than our own. Thus, by DNA analysis, we humans are truly part of one family." "Population geneticists... look at these facts about the human genome and conclude that they point to all members of our species having descended from a common set of founders, approximately 10,000 in number, who lived about 100,000 to 150,000 years ago."

Controversy will continue, but a new understanding of a fully integrated, living universe is emerging. It is finally undeniable that human beings also arose as an integrated part of the whole spectrum of life. The ancient quote "As above so below" is taking on new and clearer meaning. The idea of being "in the image of God" begins to take on new significance. The fact that humans are so different from all other creatures on earth makes sense if we are adding a new dimension and new possibilities to organic life on earth.

However, some scientists are atheists, so we must give them their due. They look at the evidence and see only accident. Some insist there is no creator or purpose to life, no meaning in existence, and certainly nothing survives after our death. For those who believe in God, these ideas are even difficult to hear. But believers also need to recognize the difficulty atheists have in hearing about a personal and loving God. But whether we like it or not, our ultimate hope lies with each other.

And we all have to accept that the final word has not been heard yet. We cannot create a single living cell from raw materials. Much remains to be discovered. How strange humans are that upon seeing half a glass of water some will say its half empty and others will say it's half full. What is the nature of reality? Are human beings incapable of living in harmony with one another? Is our current stage of development unfinished and incomplete?

So what do people find so objectionable about the Theory of Evolution? First, that some supporters of evolution seem most interested in proving that a God or Creator does not exist. Secondly, that the idea of "the survival of the fittest" is a cruel and unforgiving notion. And thirdly, the image of an all knowing, all powerful, loving and personal God seems contradicted by all the chaos, accident, and untimely, unjustified death, disease, and suffering in the world. This is the image of a failed God that the atheist Victor Stenger writes about in his book "God: A Failed Hypothesis".

Each one of us is responsible for our view of life. We are all different and unique as human beings, and there will always be different points of view. Our point of view is to recognize the truths and validity contained in both the creation and the evolutionary perspective. To seek God and to seek truth has the same meaning for us. God is truth, but beyond our grasp, so we need to have faith. And we need science to continue its search to discover the laws of nature, and to continually make glorious the unknown wonders of our universe.

If you read both Francis Collins and Gerald Schroeder you will learn about all the evidence that points to the existence of some kind of creative force that must exist behind everything we see in the universe. What that creative force is, and exactly how you believe, and what you believe, has to be left for your own search.

On the other hand, the increasing degrees of consciousness revealed in the pattern of life form development, the possibility for human beings to reach a still higher level of development we call spiritual, the ideas of different levels of energy, and the unified wholeness of the universe are intended to provide a perception that connects us with our universe. We know the cells in our body are connected to form a human being and that humans are cells in the tapestry of organic life on earth. Just as the living earth is a cell in a cosmic framework we are only beginning to understand.

Our universe is one where matter and energy is everywhere in movement between the extremes. Just as God is truth, God is also the highest form of energy. Through evolution matter is organized to become life. Life provides new energy. A part of that energy returns to the source, and a balance in the universe is maintained. The teachings of spirituality and science help each of us strive toward a richer, deeper participation in this mind-boggling experience we call life.

In the coming years we hope more scientists will write books proposing a view similar to that of Collins and Schroeder, and that there will be many more books and discussions like ours, attempting to make these ideas more available and open to a more spiritual perspective.

The Possibility of another Level of Development
We Call Conscious or Spiritual

The physical and spiritual realms always go together. Spiritual existence is contained within physical existence. Spiritual existence derives its energy and growth from food digested and transformed in the physical realm by living cellular creatures. However, while physical existence is visible, spiritual existence is much more difficult to recognize and define. Many people simply deny the existence of any reality besides the one we see and experience every day. Others indulge themselves in a spiritual reality made up of their own interpretations, hopes, beliefs and fantasies. In this kind of spiritual world everything and anything is possible and is limited only by one's own imagination.

Thus many popular beliefs are based upon other worldly forces that magically intercede into this world, while they simply ignore the concept that all forces belong to one and the same universe, and thus obey the same laws, which is consistent with Big Bang Theory and our approach.

Many people believe man controls his own destiny while others believe God is in control. We bring the viewpoint that spiritual development is an integrated aspect of evolution that requires the harmonious functioning of our physical, emotional and mental faculties. Certainly our basic abilities are needed for physical survival, but to evolve spiritually, we need to become aware of our connection to the whole of ourselves, which represents a distinctly different direction from physical development alone. Spiritual development is independent of the physical survival instinct.

The fact that so many human beings do not seek or believe in anything they cannot see and touch with their own senses, is evidence that spiritual development is not something shared by everyone, while everyone certainly does share a physical survival instinct. Those who interpret Darwin's theory as denying the spiritual realm, leave us with a theory cut off from a significant part of the human experience. And when the survival instinct is interpreted as being something purely physical, the theory can no longer account for the entire human experience. Darwin's theory is at a similar level of truth as Sir Isaac Newton's theory of gravitation, and just as Newton's theories were expanded and superseded by Einstein's theory of relativity, so now we await the appearance of a theory of evolution that upgrades Darwin's concepts to include the phenomenon of consciousness. Neither Darwin nor Newton felt their own theories excluded a creative force existing above and beyond what our senses perceive. It took smaller minds to reach such conclusions.

So let us examine the continuum of existence, where both the physical and the spiritual worlds exist together, within us, as well as outside of us, in the propagation of matter and energy we call the universe. Plants combine the energy of the sun, nutrients, water from the soil, and carbon dioxide to live. Does a plant have consciousness? A plant's leaves will turn and follow the sun. Isn't this a form of consciousness? Animals combine nutrients from plants, other animals, oxygen, and water to live. They nurture their young, move with the rains and the seasons to find food. Isn't this a level of consciousness? Each form of life combines matter and energy and is conscious of something, but the materials used and the levels of consciousness are different. Humans combine plant and animal nutrients, water and oxygen to live. But of what are we conscious? This is where we seem to be potentially different from animals. Some of us live just as animals live, and so for those individuals the level of consciousness must be similar to that of animals. But why do we have myth, music, art and science? We seem to have evolved higher capacities, and that suggests our evolution is not limited to the physical realm alone. We have feelings of awe, amazement and joy. We even sometimes sense a connection to something difficult to explain or put into words. Why is this so? Our will to search for more than just food and shelter is evident. How is our evolving free will influencing us? Is something driving our consciousness to evolve? Could an evolving consciousness be the way humans are to participate and serve in the continuum of life?

As a practical matter, none of this is relevant to us unless we have a wish to search in the spiritual realm. What is the meaning of what Jesus told us, that: "The Kingdom of Heaven is within you; and whosoever knoweth himself shall find it". Why should we "love thy neighbor as thy self", or live by the later Christian concept, the golden rule: "do unto others as you would have others do unto you"? George Bernard Shaw's saying that "the only problem with Christianity is that it has never been tried" is just a clue that we don't understand our fundamental faculties. We don't function harmoniously. Jesus said "He who will drink from my mouth will become as I am: I myself shall become he, and the things that are hidden will be revealed to him." So, if we can begin to find harmony within ourselves, we can strive to find harmony with what is outside as well, and begin to evolve as individual human beings.

As we become more sensitive and develop along a continuum to higher levels of capability, which might include our everyday job survival skills, or even driving a car, the boundaries of the physical, emotional and mental realms begin to harmonize, and a spiritual realm of perceptiveness emerges. Being an expert or having expertise reflects this spiritual realm. However, the expertise of most people is limited to their area of specialty, and their spirituality is not often reflected in other aspects of their lives. In their area of specialty they may have a greater degree of harmonious development, but with respect to their whole being they may still remain quite ordinary and undeveloped.

True spiritual development is aimed at the whole individual, and is related to a deep experiencing of the whole of life. During our lifetime our physical body manifests chaotically between one level and another in the instinctual, sexual, moving, emotional, mental and spiritual realms. Thus our worldly image is created. As we become increasingly conscious of the truth of how others see us, which is the equivalent of knowing ourselves, we become more able to move in the direction of greater harmony. In fact, this greater awareness is what empowers us to move in any direction we so desire. But please don't think for one moment this is easy to accomplish, it isn't. An important analogy, and perhaps also a warning, would be like taking a miracle medicine to cure something and ignoring the side effect warnings on the bottle. Western medicine reflects the exact same attitude as people who think they can change themselves. This attitude assumes they know and understand enough to make these changes. The fact that dangerous and even deadly side effects often result is proof positive that they do not know or understand enough. So we must be very cautious about changing the delicate balance of things. Another difficult aspect of wishing to change is that many of our traits do not need to be changed. To function in the everyday world we need our good old personalities, or at least most of it. So the real change is of a very subtle nature involving one's inner self awareness. How it may manifest externally is hard to judge or evaluate.

    The spiritual realm is indeed invisible because its function and usefulness to the universe often lies in a dimension beyond our ordinary, disharmonious perceptions. This is why in order to properly evaluate an aspect of the spiritual world one may need appropriate training. And why an untrained eye sometimes jumps to the wrong conclusions. This is what makes spiritual manifestation so vulnerable to misinterpretation.

The spiritual realm consists of a relationship of forces that unify a lower level with a higher one. Each level operates at a different speed and is made of different materials. We can recognize the differences between the speed of physical movement, the velocity of an emotional response, and the moment of a flash of insightful thought. Our everyday actions and manifestations represent the sum of the level of our existence. Each one of us has to ask the question, am I doing the best I can? Am I satisfied with how I am?

The spiritual realm is only sought by those, who of their own free will, feel and believe a more harmonious state is possible. This higher possibility has significance for their aspirations in life. No one can make us believe in the spiritual. We need our own inner vision and experience of it.

Those scientists who conduct experiments and base their conclusions on the assumption that all matter exists on the same level ignore the effect of harmonizing the energies of different levels. They may try to explain brain functioning based on electrical impulses coming from nerves. And indeed we have already learned much from recognizing the electrical nature of our own internal communication system. But to believe the positron scans and other measuring devices are describing the whole picture is naïve. To expect to answer questions involving spiritual matters with this approach alone is misguided. And further, to create whole fields of science based solely on this methodology is deluded. This reductionist approach may explain some of the trees, but it is doomed to forget the forest. All high levels of human skill involve a great degree of harmony between the basic elements of body, mind and feeling. Take any successful musician, athlete, artist, or scientist as an example. Even though they don't necessarily represent a high degree of spiritual development, nevertheless we do respect and even admire their superior levels of skill. It is precisely because no other creature on earth can attain sufficient harmony between these three elements that the capacities of human beings are unrivaled on Earth. Even with today's technology, and throw in tomorrow's as well, mankind will never reach heaven on the tower of Babel.

Human beings are a life form capable of harmonizing physical, emotional, and mental parts into a whole and spiritual being. But in the spiritual realm, the required harmony does not seem to have taken place yet for the mass of humanity. We know of at least 10,000 years of human history plagued with disharmony and violence. The only exceptions are famous individuals and certain small groups. Does this imply that the defining characteristic of the spiritual realm is that it can only be attained through conscious intention on the part of specific individuals?

Or in other words, might it be true in the spiritual realm, that evolution, having provided the possibilities and raw materials, requires the help of its own creation to reach the aim of its own existence? And was it for this subtle potentiality that nature or evolution was, indeed, created?

Evolution in our own Lifetime

Based on the current level of personal and social behaviors, we cannot yet consider humanity to be a finished product. The role of spiritual teaching is, and always has been, to remind people of their imperfections, and to guide them to take the actions needed to continue spiritual development. Imperfections are our own inner disharmonies that prevent us from approaching a spiritual level.

As physical life forms evolve, their spiritual possibilities increase. Nature evolves vertically as well as horizontally. Turtles, sharks and crocodiles, are among the oldest and most successful living creatures, but why didn't they turn into human beings? Did their successful survival mean that evolutionary progress slowed down or stopped? Why is it that creatures of higher consciousness developed later? Can we say evolution reveals a continuum of increasing consciousness?

In any case evolution has put its most intelligent, feeling, and conscious creature in charge of our planet. Still, our insect friends could take over some day. When comparing a human to a shark, we see vast differences as well as certain similarities. Both species are extremely successful but, with respect to intelligence and adaptability, humans win easily. Sharks are excellent in a narrow range of behaviors, whereas humans manifest a far broader spectrum of successful adaptations. But humans also demonstrate a range of self destructive behaviors. This does not appear to be true for insects or sharks. Except for rare circumstances other creatures are not self destructive.

Humans have become more independent of evolution and natural forces than any other creature. Might we say our adaptive survival skills have evolved into free will? Could we also say we have entered into a new evolutionary stage called the continuum of free will? Does free will introduce possibilities for mankind that are not available to other species? What does self destructive human behavior reflect about the new possibilities offered by free will?

So, at each stage of development, life seems to generate a certain quantity and quality of energy. Quantity is represented by successful survival. Quality is represented by the level of consciousness. Spiritual energy is energy created on a lower level to serve a higher level.

Upon the death of each creature, all of its matter is returned to the universe. This recycling takes place according to the energies contained within the matter of each creature. Some goes into the earth itself, some may be needed by other creatures. Some may find its way to places we don't see with our ordinary senses.

Given the possibility that some energy may reach a higher level, survival takes on new meaning, now on a spiritual level. Each continuum evolves within and serves another. A sense or striving may appear that urges us onward toward something beyond our ordinary grasp. Just as physical birth brings us into this plane of existence, spiritual birth prepares a next level of energy and perhaps existence.

All life participates in birth, living and death, but nothing spiritual takes place without a corresponding growth of consciousness. Nothing conscious happens without the participation of free will. As we become more sensitive to conscience, the application of our free will evolves to serve the consciousness continuum.

The main point of this evaluation is to begin to acknowledge the relation between body, mind and feeling and that higher levels of energy can arise out of the proper functioning of these three parts. The stability of atoms depends on the balance created by positive, negative and neutralizing forces. Our spiritual possibilities depend upon the harmony we can create among the three forces existing within ourselves.

Review of Part Two

Evolutionary studies reveal that as life forms developed through time, consciousness increased. This unfolds in significant stages. One celled creatures that first incorporated the miracle of DNA, multiple celled creatures, plants, fish, amphibians, land reptiles, mammals and humans. The functional significance of these stages or levels is that up to land reptiles, we have primarily a physical body with its senses and reflexes. Beginning with mammals we see the essential development of the emotional body. This new emotional body is capable of extending its relationship with the surroundings in a far deeper way than the physical body alone. And with humans the brain and mental faculties begin to come into play and a still greater depth of relationship with the environment becomes possible.

The modern understanding of Physics is that the universe is not completely predetermined. The universe contains uncertainty and also the unknown. Much is mechanical, all living organisms run like machines, but no living organism is the same as a clock, computer or automobile. Living organisms are more than machines. All living creatures have some degree of consciousness. Machines have no consciousness at all.

Atoms are made up of particles and fields. Each atom functions as a whole, and has the ability to interact with other atoms. Atomic interactions produce new combinations of matter with entirely new properties. So the first form of matter, non-living matter, exists on the atomic and molecular level.

The second form of matter, living matter, exists on the cellular and organ and body level. The third form of matter, spiritual matter, exists on a level above that. The level depends on the degree of consciousness and being. Today science recognizes forms of energy such as light, x-rays, gamma rays, and cosmic rays, among others. But these energy forms are not considered alive. In the future perhaps energies such as thought, feeling, consciousness, and even love might also be considered as energy rays. Might they also be regarded as alive as well? Creation began at the source of everything and spiritual life returns to the source, thus the circle of life is completed.

Just as Science must continue to explore the mysteries of the physical universe, we must continue to explore the mysteries of the spiritual realm. For human beings, this exploration is vital, and that brings us to the next subject in our book.

The Soul of Nature
Part Three

Continua and Free Will as Spiritual Developments

The development of life forms on earth follows a continuum of stages. Each stage is directly dependent upon prior attributes. New capacities are added on to attributes present in the old stage. As Gerald Schroeder says "Organization is the difference between living and non-living." Entropy, a term used to describe the natural tendency of unguided energy to dissipate, cannot by itself lead to life. Life develops in the opposite direction to entropy, opposite to the outward direction of the Big Bang. Entropy dissipates energy; the organization of life concentrates and refines it, thus energies can evolve into higher levels of consciousness.

The following represent a developmental continuum of life stages. These stages represent functional capacities rather than strictly biological or genetic classifications:

Stage	Attribute
Plants:	eat light and carbon dioxide, excrete oxygen, have no directed movement and are confined to the environment.
One-celled animals:	self directed movement originating in water.
Multi-celled animals:	relationship among cells within a larger organization.
Fish:	self directed movement and full life in water.
Amphibians:	transition to moving and breathing in air.
Reptiles:	breathe entirely in air and mostly reproduce on land.
Mammals:	live birthing, relationship with offspring, beginnings of recognition of other's needs, beginning of feeling, emotion.

Humans:	development of brain, curiosity, logic, invention, foresight, language, art, technology, music, beginning of mind.
Spiritually Developed Humans:	beginning of harmonious relationship between body, mind and feeling. Beginning of relationship with a spiritual body within our physical bodies. A sense of connectedness with the universe as a whole. Beginnings of a new understanding of the sense and aim of existence.

The life stages can also be seen as a continuum of relationships:

Stage	Relationship
One cell	within cell only
Multi-cell	cell to cell and cells to the whole
Fish	water breathing, full life cycle in water
Amphibian	transition to life on land and in air
Reptile/Bird	air breathing, beginnings of nurturing
Mammal	developed nurturing, to young, and mate relationships
Human	mind, invention, language, technology, society
Spiritual	higher relationship within self and with all that is

Each new stage can be thought of as approaching a new, higher dimension or level of relationships. Each life form is descendant from the form below it and prepares the possibility of a new relation with the level above it. Humans developed from mammals, and spiritual humans must evolve from ordinary humans.

Is the Evolution of Life Forms Leading In Some Direction?

If we accept the fossil and Genetic evidence of life form development we can no longer deny the increases in sensitivity, awareness and intelligence that continue to be acquired by life forms. It is clear that when life began the physical level was dominant. Gradually the influence of emotion grew to integrate with the physical side of organisms. The development of mind followed. With human beings today we witness the beginnings of the integration of mind with the physical and emotional sides of life. Each of these three realms, the physical, emotional and mental, adds a new dimension to the experience of being alive. We humans have certainly reached a stage where we, or at least some of us, have begun to reflect backwards, wondering where we have come from, why are we here. Can we say human beings have evolved to the point where we need to understand why we are here? With the appearance of humanity could the outward direction of the Big Bang have evolved to a turning point? Now, through us, energy has a means through which it can begin to make a return journey to its source. Is this what lies behind the meaning of a spiritual direction?

Today many believe in a personal God, but many also believe in a distant God, One who created the universe but basically leaves us alone. We submit that the concept of "free will" can help reconcile differences between these two views. While the universe contains cause and effect it is not rigidly predetermined. The universe is much more than any man made machine or computer. The universe seems to be a living organism and has needs. Just like any other life form, the universe must adapt in order to maintain itself. Free will and spiritual development act to balance entropy and uncertainty. Within the individual, spiritual forces come alive. Could it be that as these forces grow within us God becomes personal?

Can you accept that, as we are, we do not possess free will? The best example of free will is given by the sacrifice Jesus made in accepting that he had to go to the cross. "... not my will, but Thine, be done". Gospels of Matthew, 26:42, Mark, 14:36, John, 18:11, and Luke, 22:42. The complete opposite of free will is given by the example of current suicide bombers. Free will is not a matter of having enough will power to fulfill one's own desires, wishes and aims in life, or to just do what you are told. Most of the time strong will is an admirable quality, but free will involves a conscious development of the whole human being, which brings will power or free will to an entirely different level.

The only total will suicide bombers represent is ill will, along with delusion, complete gullibility, unhappiness with life, and unimaginable anger and hatred. These are not the qualities of free will. Free will in a human being can only appear when that individual has reached a high degree of independence from reacting to surrounding influences. This perspective is reflected in the expression "to turn the other cheek". The actions of such a person reflect decisions made on a plane of existence higher than our ordinary level. Free will is comparable to the action or "Word" of the creator, "let there be light", or on an earthly scale, the actions of Jesus Christ, or someone else, also far advanced along the continuum leading to a higher level of developed being.

The attribute of free will grows and develops along a spiritual path. It is a defining attribute of spiritual attainment. It has many levels and degrees, and is often difficult to recognize or accept. From the viewpoint of apprentices we are reminded that spiritual development represents a level of understanding higher than our own, which is what makes it difficult to accept. Nonetheless, it defines a true spiritual path, and also defines those individuals you meet along the way.

The development of free will is also the same attribute that makes all spiritual paths extremely slippery to travel. Staying on a path requires making decisions, and you must accept responsibility for those decisions, for it is by your decisions that you climb or fall. Your decisions, your choices, and thus your actions on a spiritual path and in everyday life as well, are developing free will all the time. Nothing is for certain, our destinies may not be predetermined, and so we always need faith to face challenges voluntarily. However, common sense is also extremely useful. Everyone has feet of clay, so we must not allow faith to blind us.

The Birth Continuum, Four stages of Birth Evolution: Cell Transference, Instinctive Reproduction, Emotional Reproduction and Soul Birth

Let us take a look at the miraculous process of reproduction or giving birth. Microscopic, one celled animals dominated the earth for over 2 billion years. They reproduced directly by a process of cell division called mitosis. Their DNA was already fully developed, and was transferred to new cells directly by what is called horizontal gene transference. When a new trait or attribute contributed to survival, most of the surviving population of one celled creatures got the new trait almost immediately, practically in one generation. This process was very instinctive and automatic. Modern viruses adapt with incredible speed to new agents designed to destroy them. The relative elegance and simplicity of their organization is largely responsible for this.

Next came the egg layers. Fish, amphibians, dinosaurs and other reptiles and birds, dominated the world for 300 to 400 million years. Reproduction was instinctive, that is, it was characterized by a relatively low level of care giving. At this stage very little learning was passed on to the next generation from the life experiences of the prior generation. Physical DNA was the dominant force affecting evolutionary change.

Mammals began to appear during the times of the dinosaurs, but remained as very minor species. Mammals only began to become dominant after the demise of the dinosaurs, around 65 million years ago. Current scientific belief is that an asteroid crashed into the earth in the area of the Gulf of Mexico and created the equivalent of a nuclear winter, thus contributing to the end of the age of dinosaurs. How evolutionary development on the earth would have been affected if the dinosaurs continued, must remain an intriguing but unanswerable question.

Live bearing mammals quickly became dominant. The greater capacity for feeling and emotion expressed by caring for and protecting the young evoked a deeper and more complex relationship between the generations and with the environment as well. While there were other contributing factors, we believe the introduction of feeling was the most significant ingredient in the eventual evolutionary success of mammals. The inclusion of emotion, in addition to the already well suited physical body, marks a third stage of reproductive evolution.

As humans began to evolve within the family of mammals, beginning about 1 to 3 million years ago, a further stage of evolutionary development emerges. The human brain begins to provide unprecedented faculties. Humans have the ability to reflect on themselves. We bring new levels of curiosity, dexterity, intelligence, self-inquiry, language and social relations. And somewhere, somehow, we begin to concern ourselves about life itself. What is life, what happens to our life energy in death? When physical movement stops, apparently nothing is left. But matter is always in motion, nothing ever really stops. What happens to the energy that was our lives?

As human beings strive to attain real answers to these last questions, we begin to enter upon a fourth stage of evolutionary development. The emergence of the human mind, made possible by the evolutionary development of the brain, allows a fourth stage of evolution to begin, the development of a soul.

Schroeder quotes Einstein on page 8 of "The Hidden Face of God" "Matter is merely condensed energy". As matter disintegrates the energy from that matter is released into the universe. It must go somewhere. Collin says on pages 71 to 73 in "The Theory of Eternal Life" that judgment among the ancient Egyptians was depicted by measuring the weight of the heart versus the feather of Maat. The heart being lighter than the feather meant the individual had reached higher spiritual development compared to when he had come into the world. "Weighing represents a judgment which is absolutely impersonal, an objective measure of density. It is the assessment of man's being by the operation of natural laws." The concept of measuring a soul also brings to mind the parable of the Talents in the New Testament Gospel of St. Luke 19: 2. We shall receive, in accordance with how we have increased that which has been given to us.

    The fourth stage of development is illustrated by examples of great and exceptional human beings such as, Moses, Buddha, Jesus, Mohammad, and certain others who are recognized without dispute, at least by most of those of their own cultures. These individuals are considered exceptional because of their behavior when compared to all other individuals of their own time and place. The standards and principles by which they lived and died, and their own personal, total devotion, to these standards and principles, marks them as unique in the hearts and minds of not only those who knew them, but also most of us, who have only heard of them.

Entrance into this fourth stage of human possibility implies the birth of a soul, which is a new body, but this time on a spiritual level, invisible at the physical level. This is the re-birth Jesus speaks of in his dialogue with Nicodemus. How the soul develops is what all spiritual teachings are about. It is well beyond the scope of this book to speak about how, but, it is the main point of our book to state that, the development of the energy we refer to as "soul", is the aim and purpose of the continuum of evolutionary forces, and what they represent. Is consciousness the energy that can evolve into free will?

Evaluating Biblical Passages from a Modern Perspective

Because of significant cultural differences it seems unfair to try to use Biblical texts to justify strictly modern issues. What is amazing, from the viewpoint of evolution, is how much of the sequence of life form development, from plants and trees, to sea creatures, reptiles, birds, mammals and finally Adam and Eve, the authors of the Bible seem to have gotten right. We need to resolve the issues of our day and age with knowledge and wisdom gained in our own time.

Genesis begins with the creation process in order to express two main ideas. The first is "as above so below". The second is "God made man in His own image". According to Genesis after God created man, He rested. God declared the 7th day, a day of rest, a holy day.

In our interpretation Genesis establishes God's creation process as a guide for the pattern of human life as well. God did His work in 6 days and rested on the 7th. According to Genesis human beings are to do the same. It is by following this proportion of devotion to earthly duties, and devotion to spiritual duties, that human beings can fulfill God's plan. His intention is for us to be in His image. To be like God is to behave like God.

God created everything and rested only after He created man. Humanity is the ultimate fruit of creation. Just as God creates the universe for mankind, humanity must create spiritual energy for God. All of life, all of existence, is a balanced exchange of energies, the great circle of life. After all the preparatory work, God created mankind, then He rested, it is up to mankind to do the remaining work.

For those to whom it might matter, Gerald Schroeder provides a scientific calculation that reconciles the Bible's 6 days of creation with today's estimate of a 15 billion year old universe. According to Schroeder both views are correct. You can find this on the internet at: http://www.geraldschroeder.com/age.html.

The following is a synopsis of Schroeder's reconciliation of the 6 days of creation in Genesis and the 15 billion year old beginning of the universe according to Big Bang Theory: Scientists calculate it took about 15 billion years for the universe to reach its current size. They also calculate that the relationship of time around the beginning of the universe to time today is about a million million to one. If a light was flashed once each second, starting when the universe began, how long would it take the first flash to reach us today? Answer, one million million seconds. If creation took 6 days to reach the creation of Adam, that would be 6 million million days. If we divide that by 365 we get approximately 16 billion years.

The 6 days is looking forward from the creator's perspective, while from our perspective looking back into the past we have 15 billion years. A creator's time cannot be measured by our time. A creator's time would move much more slowly than ours, our guess is perhaps a million million times more slowly.

Seven as a Sacred Number

There is even further natural wonder in the creation theme of 7 days.

We are all familiar with the 7 notes of the musical octave, which is recognized by almost all cultures in the world as a musical standard for the most fundamental pattern of musical repetition. The 7 colors of visible light are another example of the natural repetition of processes by sevenths. (red, orange, yellow, green, blue, indigo, violet).

Even arithmetically, when you divide one whole into seven parts you get:
1/7 = 142857142857142857142857142857...

The value $1/7^{th}$ is an endlessly repeating sequence of the same numbers in the same order, reflecting the infinite nature of a single pattern repeating itself by sevenths.

2/7 = 0.285714285714285714...
3/7 = 0.428571428571428571...
4/7 = 0.571428571428571428...
5/7 = 0.714285714285714285...
6/7 = 0.857142857142857142...

It is no coincidence, infinite wisdom lies behind the choice of a seven day creation.

Turning the Creation Process Inward

Our book has been all about trying to illustrate the spiritual realm. We need to distinguish our everyday experience of the world from another possible range of experiences which appear only as rare flashes or glimpses. It also happens that many of our most spiritual experiences come from unknown sources and happen totally unexpectedly. However, prayer and meditation involve our attempts to bridge or relate these two realms of experience, the every day and the rare or esoteric. Whether we conceive of the higher level we are trying to reach as being outside ourselves or within ourselves or both, we can all acknowledge the reality of degree, or of continuum. Seeing the contradictions that exist within us is the key to real search. In our present disharmonious state, the moment we stop seeing these contradictions, any prayer or meditation we are doing becomes a daydream. It means the connection is broken. The experience is no longer sincere. You are no longer present to the actual experience taking place in the moment. To find, recognize and sustain this connection with the dynamic present moment is what we need to practice. The rest is not up to us.

Now let's try an experiment where we see if we can experience the same lawful process of creation, only now this process continues inside our own bodies, now an inner creation. I see it is not so easy to make this change of attention an inner reality. My mind, the attention, I'm confused I don't know what to do, how to be. How do we turn the whole attention inside?

Direct the mind toward the body, bring the consciousness inside, and become aware of myself, my body, my thoughts, and my feelings. I become calmer, quieter, I sense myself as more present, more here. Now that I am making this effort to bring the attention within myself I become more aware of what is taking place, of where I am, of how I am. Yes, I can connect with my own consciousness, even though only fleetingly and incompletely. Nonetheless, this experience gives me a glimpse of something new, something alive.

This exercise represents the process of creation continuing within us. Could this exercise help us understand what generates our soul matter and soul energy? It also helps me understand why at first I cannot connect. I see that my attention is not wholly and consciously involved in the process of ordinary activities. My mind, my feeling, is all over the place, disharmonious. A conscious and whole attention is what is lacking. The difference between ordinary people and those who are trying to develop is the willingness to accept this condition, as it is, and to allow their search to begin there. Only humans have the unique ability to awaken to their own consciousness within.

The Messenger Evolves

From the beginning the word of God has been delivered through messengers. Is the image of God evolving? Can we see the idea of the messenger evolving? First, through the Bible we see God Himself doing His own miracles, then He gives His laws to Moses, later He sends Jesus to forgive and reconcile, then Muhammad brings another reminder to pray often and not to forget. As religions wander further away from their source, divisions multiply. Eventually a new form of search begins to evolve, called science. Initially science believes it will be able to explain everything, even prove that the universe is nothing more than an accident. This generates even more divisions. But soon science begins to discover the universe is uncertain. There is a something unknowable, beyond our reach. The search for truth changes what is unknown. The search for truth changes us. Scientists, religious people, and spiritual people, begin to find common ground. Kings, Queens and Emperors can no longer rule alone. People begin to demand democracies, parliaments, voice in government. The human search for truth begins to include more and more people, mutual respect guides our way back toward unity. We can all strive to become messengers.

Summation

It is very easy to criticize ourselves for all the mistakes we have made in our attempts to create civilizations, but perhaps it has all been necessary and unavoidable. We must learn to understand our past. The movement of evolution towards increased independence and increased consciousness emphasizes the growing significance of each individual within the population. The role of the individual is rising and evolving.

Three hundred years ago we had to resolve the issue of earth revolving around the sun. Today we have to resolve the question of creation in 7 days and also in 15 billion years. We need to see the parallel. We need to find reconciliation. We all need to participate. We all need to move toward truth.

The Bible and other traditional spiritual texts are living testimonies to the presence of great wisdom in ancient times. In Genesis 4-6, 7 "And the Lord said to Cain, Why are you angry? And why is your face downcast? If you do right, will you not have honor? And if you do wrong, sin is waiting at the door, desiring to have you, but do not let it be your master." Thus, the Lord speaks to us all, and we need to strive to comprehend this meaning.

Later, after Cain kills Abel, (Genesis 4, 10-12) God gives Cain only two direct punishments: First that when he works the ground "it will no longer yield its crops for you", and then "You will be a restless wanderer on the earth." Our interpretation is that without being in God's presence, we are doomed to be restless wanderers on the earth. Unless we seek spiritual development, we will not rise above the level of earthly existence.

Growth and development are natural attributes and properties essential to all forms of life. But human life has a unique imperative to develop. Current scientific evidence as well as spiritual traditions indicates all of known existence has one common point of origin, and follows one set of laws or principles. There are different levels of consciousness within the plant and animal world, within organic life on earth, and perhaps beyond this in a manner we have yet to fully understand.

"The Soul of Nature" has regarded human beings as distinct and unique among living creatures because of the possibility of functioning harmoniously in body, mind and feeling. Harmonious functioning develops the spiritual energies, which otherwise will remain only as potential.

Bibliography

Anonymous *The Cloud of Unknowing* Edited by Evelyn Underhill Dover Publications, Inc., New York 2003 (from original material published in 1912)

Bennett J. G. *Talks on Beelzebub's Tales* Samuel Weiser, Inc. 1988

Collin R. *The Theory of Eternal Life* Vincent Stuart Publishers Ltd. London 1956

> (Pg. 31) Light travels instantaneously in three directions, that is, not only along a line like a cellular body, nor over an area, like a smell, but throughout a volume of space.

> (Pg 71) Weighing represents a judgment which is absolutely impersonal, an objective measure of density. It is the assessment of man's being by the operation of natural laws.

> (Pg 73) This heart is weighed against the feather of Maat, that is, right, truth, and particularly what the individual should be, his right way or true potentiality. The

being that the man has evolved for himself is thus measured against his original capacity, just as it is in the parable of the talents.

(Pg 117) The author explains, in reference to using a logarithmic scale to map out a lifespan scale for all levels of the universe, that the longest lifespan is 80,000 years representing the mineral world, and 2 ½ seconds represents lifespan at the highest level, the electronic world. The proportion between 80,000 years and 2 ½ seconds is a million million to 1. Note below that G. L. Schroeder also uses this same proportion.

Collins F. S. *The Language of God* Free Press, A Division of Simon & Schuster, Inc. 2006

Cook T. A. *The Curves of Life* Dover Publications, Inc., New York (from material originally published in 1914)

Dawkins R. *The Selfish Gene* Oxford University Press New York 1989

de Llosa P. *The Practice of Presence* Morning Light Press Sandpoint, Idaho 2006

Gould S. J. *Wonderful Life: The Burgess Shale and the Nature of History* Penguin 1991

Greene B. *The Elegant Universe* Vintage Books a Division of Random House, Inc. New York 2000

Gurdjieff G. *All and Everything: Beelzebub's Tales to His Grandson* E. P. Dutton & Co., Inc., New York 1964

Hagedorn H. *Prophet in the Wilderness: The Story of Albert Schweitzer* Reader's Digest 1994 reprinted by permission of the Macmillian Publishing Company

Herz-Fischler R. *A Mathematical History of the Golden Number* Dover Publications, Inc., New York 1998

Keating T. *Open Mind Open Heart* Continuum International Publishing Group, 1994

Lawlor R. *Sacred Geometry Philosophy & Practice* Thames & Hudson Ltd. London 2005

Lipton B. H. *The Biology of Belief* Mountain of Love Productions Inc & Elite Books 2005

Malin S. *Nature Loves to Hide* Oxford University Press New York 2001

Needleman J. *A Sense of the Cosmos* Monkfish Book Publishing Co. 1975, 2003

Ouspensky P. D. *In Search of the Miraculous* Harcourt Brace & World, Inc. New York, 1949

Runion G. E. *The Golden Section* Dale Seymour Publications New Jersey 1990

Schroeder G. L. *Genesis and the Big Bang* Bantam Books 1992

Schroeder G. L. *The Science of God* Broadway Books 1997

Schroeder G. L. *The Hidden Face of God* Free Press, a division of Simon & Schuster, Inc. 2001
A view of matter from physical science:
(pg 4) "If we could scale the center of an atom, the nucleus, up to four inches, the surrounding electron cloud would extend to four miles away, and essentially all the breach between would be marvelously

empty."

(pg 6) "The solidity of iron is actually 99.9999999999999 percent startlingly vacuous space, made to feel solid by ethereal fields of force having no material reality at all." "...atoms of things we consider inert are not any different physically after they have become part of a living organized entity. Organization is the difference between living and non-living."

Schroeder G. L. *website*
http://www.geraldschroeder.com/age.html
The calculation that reconciles 15 billion years with a 6 day creation is based on the following:

The temperature of Quark confinement; the point at which matter freezes out of energy, which is 10.9×10^{12} degrees Kelvin, is then divided by 2.73 degrees Kelvin, which is the temperature of the universe today. The proportion of these two values equals a million million to 1.

The temperature of the universe dropped to this point (10.9×10^{12} degrees Kelvin) in a very short time after the Big Bang took place. At this point hydrogen matter began to be created and the universe began to expand. Time and space can both be said to be measurable from this point onward. As the universe expands it creates space or space stretches. The larger the universe becomes the further apart each particle moves from another.

Current science believes light travels at the maximum possible speed in the universe. Thus a photon of light emitted at the beginning of space/time would take one million million units of space/time to reach us today. If creation took 6 days, according to the expansion of space/time, that would be the equivalent of 6 million million days today. If 6 million million days is divided by 365 we get approximately 16 billion years. According to the expansion of space/time the first day lasted 8 billion years, day 2 lasted 4 billion years, day 3, 2 billion, day 4, 1 billion, day 5, ½ billion, day 6, ¼ billion years.
Matching this with cosmology, paleontology and archaeology, modern science and the Bible are amazingly close.

Schwaller de Lubicz I. Chick-Pea Herbak
Penguin Books, Inc. 1972
(Pg. 331) "That is right: the bank (of the river) offers an obstacle to the flow of the

water. It causes it to flow back, and thus a current in the opposite direction is formed. This shows the law of *reaction* and gives you, O my son, the first lesson of Wisdom."... "Every natural cause has an effect; this effect is the direct consequence of this cause. If you judge the facts according to their appearance, you will be deceived as to the true workings, and your reasoning will be erroneous.

In reality the effect is always indirect, in the sense that the cause must be thrown back by a resistance of the same nature; this will provoke a transformation of the two forces, and this transformation will give birth to the effect. It is thus that the seed acts on the substance of the ovule, and the two will annihilate each other so as to give life to a new being."

Schwaller de Lubicz R. A. *The Temple of Man* Vol 1 & 2 Inner Traditions International Ltd. 1998

Stenger V. *God: The Failed Hypothesis* Prometheus Books, 2007

The Holy Bible New International Version, Zondervan Publishing House 1989

www.ingramcontent.com/pod-product-compliance
Lightning Source LLC
Chambersburg PA
CBHW031254290426
44109CB00012B/582